Francisco W. Alcântara de Freitas
Janiquelle da Silva Rabelo
Valsergio Barros da Silva

Floristic and phytosociological survey of the caatinga

Francisco W. Alcântara de Freitas
Janiquelle da Silva Rabelo
Valsergio Barros da Silva

Floristic and phytosociological survey of the caatinga

Vegetation of the Córrego do Quixinxé Settlement, Ocara - CE

ScienciaScripts

Imprint
Any brand names and product names mentioned in this book are subject to trademark, brand or patent protection and are trademarks or registered trademarks of their respective holders. The use of brand names, product names, common names, trade names, product descriptions etc. even without a particular marking in this work is in no way to be construed to mean that such names may be regarded as unrestricted in respect of trademark and brand protection legislation and could thus be used by anyone.

Cover image: www.ingimage.com

This book is a translation from the original published under ISBN 978-3-330-76589-4.

Publisher:
Sciencia Scripts
is a trademark of
Dodo Books Indian Ocean Ltd. and OmniScriptum S.R.L publishing group

120 High Road, East Finchley, London, N2 9ED, United Kingdom
Str. Armeneasca 28/1, office 1, Chisinau MD-2012, Republic of Moldova, Europe
Managing Directors: Ieva Konstantinova, Victoria Ursu
info@omniscriptum.com

Printed at: see last page
ISBN: 978-620-8-53674-9

SUMMARY

The aim was to get to know the floristic composition and phytosociology, obtaining data to compare with another area of caatinga (at a different stage of succession), and subsequently obtaining information to better understand the successional characterisation of caatinga environments. The vegetation survey was carried out in an area of 1.0 hectare in the Córrego do Quixinxé settlement, which covers 5,561.63 hectares and is located in the municipality of Ocara - CE with geographical coordinates 04°36'49.8" South Latitude and 38°36'51" West Longitude. The area was divided into 50 plots measuring 20m x 10m. After this division, the plots were drawn completely at random. The phytosociological survey was carried out in the same order as the plots were drawn. In each plot, live individuals with a stem diameter at ground level > 3 cm and a total height > 1 m were measured. Cacti, bromeliads and lianas were not included. Twelve species were recorded, belonging to 11 genera and 7 botanical families, totalling 702 individuals with basal area = 3.47 m^2/ha and DA= 702 ind/ha. Shannon's diversity index (/-/') was 1.40 and Pielou's equability (J') was 0.54, both considered low. The species with the highest value of importance (VI) and most prominent in relation to all the variables analysed was Croton floribundus with 453 individuals distributed in all the plots, a frequency of 100% and absolute dominance of 0.92 m2/ha, which generated a value of importance of 112.40, corresponding to 90.85% of the area. Thus, any attempt to characterise the structure of this community is strongly influenced by the presence of this species, which is indicative of disturbed vegetation and early stages of ecological succession.

Keywords: Diversity Index. Species Richness.

Environmental conservation.

SUMMARY

CHAPTER 1

INTRODUCTION

Phytosociology involves the study of all phenomena related to plant life within social units. It portrays the complex of vegetation, soil and climate.

Biodiversity conservation is currently one of the biggest challenges, given the high level of anthropogenic disturbance to natural ecosystems in Brazil. In this context, studies into the floristic composition and phytosociological structure of forest formations are of fundamental importance, as they provide subsidies for understanding the structure and dynamics of these formations, which are essential parameters for the management and regeneration of different plant communities.

The peculiarity of the region's climate and, above all, its unpredictability make most attempts at subsistence through agriculture and livestock impossible without having to resort occasionally and temporarily to extractive activity. This activity is threatening the caatinga, despite the fact that it has a certain resistance to anthropogenic disturbances, such as slash and burn processes, which are systematically applied in many areas of its domain. It is well known that burning causes a drastic reduction in the volume of vegetation, as well as in the density of species in the areas thus exploited (LEITE, 1999).

Vegetation is one of the main renewable natural resources and can be used by man to harvest wood, provide shelter and food for wildlife and protect the soil from erosion caused by rain. Due to this multiplicity of vegetation functions, there is a complexity of utilisation techniques, based on the precise description of its physiognomy and floristic composition between plants and the environment (FREITAS et al., 2007).

One of the major problems for the regeneration of the Caatinga is that the population of some species regenerates very slowly and does not keep pace with the rate at which they are exploited. Some phytosociological studies have revealed that many sites sampled after logging are mostly represented by just one or a few species (SAMPAIO et al., 1997). Therefore, it is understood that some species may not regrow, their seeds may be absent and their seedlings may not become established, i.e. there is no guarantee that they will remain in the vegetation (ALVES JUNIOR, 2010).

Knowing the behaviour of native species after logging is an important tool for assessing regeneration and the success of forestry projects. Understanding how the alterations caused to native vegetation affect its regeneration helps to correct possible mistakes made during logging and ensures a sustainable yield. Therefore, the study of natural regeneration serves to assess the length of the felling cycle, which is the interval between successive cuts in a given area, and the intensity of felling, which is the number and size of trees to be removed, if appropriate, given their economic value and sustainability objectives (BUONGIORNO; GILLESS, 2002).

The system of exploitation used in the caatinga to carry out subsistence farming consists of removing all the vegetation cover (removing the usable woody material and burning the stubble). This type of agriculture has as its main objective the production of food to guarantee the survival of the farmer, his family and the community in which he lives, in other words, it aims to meet the food needs of rural families. It's worth pointing out that this system only refers to annual crops, while other areas on the same farm can be used permanently for perennial crops and pasture. This issue is part of the problem of agricultural intensification, in this case of annual crops, mostly food crops.

The usual system of shifting cultivation can be followed by successful natural regeneration. On the other hand, natural regeneration can be complemented by the seed bank or seedling bank already existing in the area. In the case of regrowth, the reserves accumulated in the underground systems and the base of the stem are expected to be sufficient to allow the plant to survive (SAMPAIO et al., 1998). However, the plant may fail to produce new shoots due to insufficient reserve storage or because it cannot withstand a prolonged period of water deficiency during the dry season (FIGUEIRÔA et al., 2008). It is hoped that the species that have been slashed and burnt for one to two years and left fallow for five years in the study area will be able to regenerate and reach a new state of equilibrium. In view of the above, the general aim of this research was to carry out a floristic and phytosociological survey of a regenerating area in the caatinga under slash and burn for one to two years, left fallow for five years. The specific objectives were to answer the following questions:

a) Which are the most abundant species represented by their importance value index?

b) What is the value of the diversity and equitability index of the regenerating area?

CHAPTER 2

LITERATURE REVIEW

2.1 Dry forests

Dry forests are those with a well-defined dry and rainy season (MURPHY; LUGO, 1986; GENTRY, 1995). These formations occupy a large part of the planet's tropical and subtropical lands (QUESADA et al., 2009). Because their average annual temperature is above 17°C, they occur in ice-free areas and evaporation is greater than precipitation for a significant part of the year (VIEIRA; SCARIOT, 2009). It has less vegetation cover and basal area than other tropical forests, and thorny and succulent species are often common, especially in the drier formations (MURPHY; LUGO, 1986).

The forests present in these environments represent the most threatened ecosystems in tropical regions, because a large part of their areas have been converted for agricultural, livestock and urban expansion uses (QUESADA; STONER, 2004; MILES et al., 2006) and continue to be one of the most sought-after resources by humans for colonisation, development and production environments (JANZEN, 1988; KHURANA; SINGH 2001; SANCHEZ-AZOFEIFA et al., 2005). Despite this, it is one of the least studied tropical habitats in the world (STONER; SANCHEZ-AZOFEIFA, 2009). An important factor about these environments is that they have unique biodiversity and a high degree of endemic species (TREJO; DIRZO, 2002), as well as being characterised by a great functional diversity of species, especially in terms of leaf strategies (EAMUS, 1999), such as deciduous, i.e. species that lose their leaves in the dry season.

In these forests, productivity peaks during the rainy season when the organic matter that accumulated in the soil during the dry season decomposes and increases the availability of nutrients in the soil (PENNINGTON et al., 2009). Tropical dry forests have markedly lower primary production than humid forests, because although both grow similarly in the rainy season, in dry forests there is a sudden decrease or even stagnation in growth during the dry season (EWEL, 1980; MURPHY; LUGO, 1986).

One of the characteristics of these forests is that even though they have a lower growth potential than humid forests, they are able to recover more quickly due to their relatively simpler structure (EWEL 1980; MURPHY; LUGO 1986; KENNARD, 2002).

Tropical dry forests have a climate similar to that of savannas, and the two can coexist in close proximity. However, savannas are wetter, have fewer succulent species (PENNINGTON et al., 2000) and are more abundant in grasses (MOONEY et al., 1995).

2.2 Caatinga

In the Caatinga, more than 50% of the trees and shrubs are medium to small and lose their leaves in the dry season. In addition, the vegetation is characterised by being rich in thorns, interspersed with cacti and bromeliads (VELOSO, 1991; ANDRADE-LIMA, 2007).

According to Pennington et al. (2000), results from recent studies suggest that the origin of the Caatinga is part of what was a wider distribution of seasonal tropical dry forest that occupied large areas of South America in dry and cooler periods during the Pleistocene.

Despite their geographical proximity in some places and characteristics in common with the Cerrado, such as the xeromorphic aspect of its vegetation, there are notable differences, mainly because it has a typically savannah-like vegetation formation, with a large presence of a sparse shrub layer with suberous, branched and twisted stems with a large presence of leathery leaves and another herbaceous-grass layer that is generally vast and continuous (EITEN, 1994; RIBEIRO; WALTER 2008). In addition, during the dry season, most of the grasses are inactive and most of their dry aerial biomass dies off, favouring the occurrence of fires (EITEN, 1994).

In the Tupi language, the term "Caatinga" means "clear forest" (LEAL et al., 2005), referring to the predominant vegetation, which is greyish and clear during the dry season, when most of the trees and shrubs are leafless and light can reach the ground.

Despite the water deficit, the semi-arid northeast of Brazil is considered to be the wettest on the planet. However, rainfall is concentrated in a short period during the year, and evaporation in some places is greater than precipitation (MALVEZZI, 2007).

In this region, where the Caatinga occurs, which is also known as the semi-arid northeast of Brazil, there are around 22 million people, representing 46 per cent of the northeastern population and 13 per cent of the Brazilian population. It is one of the most populous dry regions on the planet and where living conditions are the most precarious (MALVEZZI, 2007), making life in the Caatinga difficult for the people who live there, who are at the mercy of drought-related phenomena.

On the other hand, around 25 per cent of the energy consumed by the industrial and commercial sectors in this region comes from forest biomass, mainly from the Caatinga (GARIGLIO et al., 2010).

The exploitation of vegetation generates around 900,000 direct and indirect jobs (GARIGLIO et al., 2010). And because of this great relationship with income generation for the resident population, anthropic pressure on vegetation has also intensified over the years, mainly due to the indiscriminate felling of native tree/shrub species (DRUMOND et al., 2008), for the use of wood as stakes, firewood, charcoal and handicrafts, in addition to the various products provided by vegetation such as edible fruit and fodder (MENDES, 1997).

Despite the exploitation of its resources, the vegetation of the Caatinga shows enormous variability in terms of species composition and phytogeographic, phenological and physiognomic aspects, due to the great territorial expansion it occupies (RODAL et al., 1992; SAMPAIO, 1995).

Non-timber forestry production (fruit, seeds, bark and wood extractives) is also an economic alternative for a considerable portion of the rural population living in the semi-arid region, especially women, and in some cases it is the family's main activity.

However, these activities, carried out without proper management, contribute even more to the process of degradation of the Caatinga, mainly through the permanent use of fire (GARIGLIO et al., 2010).

Although much of the biological heritage present in the caatinga is found nowhere else but in the north-east of Brazil, little attention has been paid to conserving this vegetation and it has been put on the back burner when policies for the study and conservation of the country's biodiversity are discussed (SILVA et al., 2004). There is therefore a great shortage of information on the biological dynamics of the ecosystems within it (PEREIRA, 2001).

Given this lack of information on the vegetation of the Caatinga, the search for scientific information becomes evident, especially with regard to the processes of ecological succession and natural regeneration of the ecosystems found there, since understanding the dynamics of natural regeneration in forest ecosystems makes it possible to estimate population parameters, which are essential for sustainable forest management (DRUMOND et al., 1996).

Phytosociology is therefore an appropriate way of seeking initial answers about the organisation of vegetation and has proved to be an important tool for characterising plant communities. Knowledge of vegetation contributes to understanding the functioning of plant communities in a region.

CHAPTER 3

MATERIALS AND METHODS

3.1 Study area

The study was carried out in an area of the Córrego do Quixinxé/Antônio Conselheiro settlement. The 5,561.63 hectare property is located in the municipality of Ocara, Ceará. It has the capacity to serve 200 families (Table 1).

Table 1. Population of the Córrego do Quixinxé Settlement.

AGE GROUP	TOTAL
00 a 06	96
07 a 12	125
13 a 18	105
19 a 60	327
>60	47
TOTAL	**1200**

Source: COPASAT, direct survey - 2011.

Agriculture is focused on subsistence crops such as maize, beans and manioc, grown in a conventional system. Early dwarf cashew crops are more widely exploited, since cashews and their derivatives are the main source of income for the settlers. In addition to the above-mentioned crops, there are also various vegetable gardens from the productive backyards.

The vegetation cover on the land and also in the legal reserve areas is made up of xerophytic caatinga vegetation. This includes the presence of jurema, sabiá, marmeleiro, mufumbo and pau branco, as well as noble vegetation (wood considered to be hardwood), such as angico, aroeira, pau d'arco, jatobá and others. There is also capoeira, which after several years of agricultural exploitation has become commonplace.

The local fauna is typical of the Caatinga and includes: preá, gambá, cururu toad, agouti, armadillo peba, raccoon, fox, cassaco, chameleon, sabiá, turtledove, avoante, carcará, snakes and the rooster, among others.

The legal reserve area is 20% of the total area, i.e. approximately 1,112.2 ha divided along the length of the land,

They are duly recognised and preserved in accordance with the recommendations of current legislation. As for the permanent preservation areas such as the riparian forests, they are well devastated, since for years they were planted with swiddens and the wood from these areas was extracted. It was only when the settlers became aware of the importance of these areas that commercial exploitation ceased.

The soils are mainly made up of Argissolo vermelho amarelo, eutrophic and dystrophic, Planossolo solódico and Neossolo distrófico (INCRA/2000), which are suitable for subsistence crops, fruit growing (mainly cashew), cucurbits and extensive livestock farming.

The red Argissolo soil shows great variation in morphological and analytical characteristics, with a clayey to very clayey texture, at varying depths, with or without the presence of pebbles and gravel. Stones can occasionally be found in some of these soils.

3.2 Data collection

The area was chosen for its use in conventional agriculture, clearing it to plant maize combined with beans for one to two years, then left fallow for five years, in order to survey the composition and structure of the tree component. The pre-established criteria for characterising the area of caatinga under forest extraction were based on information gathered from local farmers. The area of caatinga under forest extractivism was defined as that which had been without agricultural activity for five years and where selective timber extraction occurs.

An area of 1.0 hectare, with geographical coordinates 04°36'49.8" South Latitude and 38°36'51" West Longitude, was divided into 50 plots measuring 20m x 10m. After this division, the plots were drawn completely at random. The phytosociological survey was carried out in the same order as the plots were drawn. In each plot, live individuals with a stem diameter at soil level > 3 cm and a total height > 1 m were measured (Rodai et al. 1992). Cacti, bromeliads and lianas were not included. Individuals were identified in the field by their common name. The synonymy and spelling were based on consultations of specialised literature.

Figure 1 - Study area

Source: Prepared by the author

Figure 1 - Study area
Source: Prepared by the author

3.3. Determining the variables

The phytosociological variables were analysed using the Fitopac 2.1 software. The following were evaluated:

3.2.1 Absolute density (AD = n/ha):

expresses the total number of individuals of a given species per unit area. Where:

DA = Absolute Density

n = total number of individuals sampled from each species

ha = hectare

3.2.2 Relative density (RD = (n/N) * 100):

Indicates the number of individuals of a given species in relation to the total number of individuals of all the species identified in the survey.

3.2.3 Absolute Dominance (DoA = Σ g/ha):

Sum of the sectional areas of the individuals belonging to the same species, per unit area. Where:

DoA = Absolute Dominance in m^2/ha

g = sectional area of each species, found by the expression:

$$g = \frac{\pi \, Dap^2}{4}$$

where:

Dap = Diameter at 1.30 metres from the ground

π = trigonometric constant pi=3.1416π

ha = hectare

3.3.4. Relative Dominance:

Indicates the percentage of the basal area of each species that makes up the total basal area of all trees of all species, per unit area: g/ha

$$DoR = \frac{g/ha}{G/ha} . 100$$

where:

DoR = Relative Dominance (%)

G = total basal area of all species found

3.3.5. Frequency:

It expresses the statistical concept related to the uniformity of horizontal distribution of each species in the terrain, characterising their occurrence within the plots in which they occur. This index can be expressed by:

- Absolute Frequency: Expresses the percentage of plots in which each species occurs:

$$FA = \frac{\text{number of plots where the species occurs}}{\text{total number of instalments}} \times 100$$

3.3.6. Relative Frequency:

This is the percentage of occurrence of a species in relation to the sum of the absolute frequencies of all species.

$$FR = \frac{FA}{\Sigma FA} \times 100$$

3.3.7. Importance Value Index (IVI):

It is the combination of the sum of the relative values of Density, Dominance and Frequency of each species:

$$IVI = DR + DoR + FR$$

3.3.8. Coverage Value Index (CVI):

It relates to the importance of the cover of each species in the environment and considers relative density and relative dominance:

$$CVI = DR + DoR$$

3.3.9. Shannon-Wiener Diversity Index:

$$H' = -\sum_{i=1}^{s}\left(\frac{n_i}{N} \cdot \ln\frac{n_i}{N}\right),$$

Where H' is the diversity index of the environment, ni is the number of individuals of species i and N is the total number of individuals considered to be the species found in the environment.

3.3.10. Pielou Equitability Index

$$J' = \frac{H'(observado)}{H'máximo}$$

Where:

H' maximum is the maximum possible diversity that can be observed if all the

species are equally abundant.

H' maximum = log S and S = total number of species

CHAPTER 4

RESULTS AND DISCUSSION

The flora of the area is made up of 07 families, 11 genera and 12 species (Table 2). The families with the greatest diversity of species are Fabaceae (5) and Euphorbiaceae (2), which together accounted for 58.34 % of the total number of species sampled. According to Sampaio (1996), these species most often occupy the top positions in floristic surveys carried out in the Caatinga, while the other families each accounted for just one species. Compared to the work carried out by (Calixto Jr. et al. 2011) in a two-hectare fragment of hyperxerophilous caatinga that was clear-cut 30 years ago and has since recovered without anthropogenic intervention, located at the Experimental Station of Embrapa Semiárido, Petrolina, PE, 432 individuals were inventoried, distributed in 08 families, 13 genera and 16 species. The families with the highest number of species were Caesalpiniaceae and Mimosaceae, both with four species, and Euphorbiaceae with three, while the others had only one representative.

Trovão et al. (2010) analysed the floristic composition of vegetation in the semi-arid region of Paraíba and also found a greater abundance of the Fabaceae, Euphorbiaceae and Cactaceae families. Representatives of these families are conspicuous in caatinga areas, as observed in other floristic composition studies (Souza and Rodal, 2010; Andrade et al., 2009; Oliveira et al., 2009).

Despite the limited influence of adjacent ecosystems as "contributors" to the total floristic composition of the areas, the phytosociological data shows a high frequency of caatinga species.

The value of the Shannon diversity index (H') was 1.40, close to the value found by Embrapa Semiárido of 1.39, whose values are in line with others carried out in caatinga environments, which range from 1.10 to 3.09 although it is considered low, partly due to the small size of the area and the history of anthropogenic intervention. The index was also similar to that found in other studies carried out on caatinga vegetation, such as Oliveira et al. (2009) in a study carried out in Serra do Monte, in the semi-arid region of Paraíba, and Ferraz et al. (2006), studying the structure of the vegetation in the municipality of Floresta, in the semi-arid region of Pernambuco. The values found are lower than those found by Miranda et al. (2000), who studied two degraded areas of caatinga in the desertification centre of Seridó, RN, and found lower Shannon index values, 1.79 and 1.86. Araújo et al. (2005), studying the diversity of herbaceous plants in microhabitats in a caatinga area in Caruaru, Pernambuco, found that the diversity, according to the Shannon-Wiever index, was 2.52. Reis (2003) found a diversity of 3.01.

The H' value of 1.40 was lower than that found by Lira (2003), 2.45 and 2.15, for anthropised caatinga environments. In another study, which aimed to analyse two Caatinga environments, the value of the Shannon-Weaver Diversity Index varied from 0.19 to 1.44, with the lowest value being obtained from an area in which 96.18% of the individuals were represented by just one species, M. tenuiflora, and the second value being extracted from an area in which A. pyrifolium and M. tenuiflora represented 77.48% of the plants sampled (Freitas et al. 2007).

The values found in this study can be considered low in relation to those obtained for other vegetation formations (Oliveira et al. 2009). However, for caatinga areas they are considerable, given the values found in other studies carried out in caatinga ecosystems, which vary from 1.91 to 3.09 (Alcoforado Filho et al., 2003; Araújo et al., 1995; Ferraz et al., 1998; Fonseca, 1991; Lyra, 1982; Rodal, 1992; Rodal et al., 1998).

The Pielou Equability Index (J') was 0.54, a value close to the 0.50 found by (Calixto Jr. et al. 2011) indicating that one or a few species predominated over the others, similar to the value obtained by Almeida Neto et al. (2009) in the Curimataú region of Paraíba. When compared to the values found by Pegado et al. (2006), who studied the shrub layer of the caatinga also in the municipality of Monteiro, PB, found high values for this parameter (J= 0.73 and 0.79), but similar to those found by Oliveira et al. (2009), studying vegetation in semi-arid regions, thus demonstrating the interspecific heterogeneity of the vegetation in the region. It is also worth mentioning that the low equability value found reflects the lower uniformity of vegetation distribution in the experimental area.

This low equability was strongly influenced by the high density of Croton floribundus, which was present in 100 per cent of the plots, indicating the low floristic heterogeneity of this shrub component. According to Sampaio et al. (1998), this species is quite tolerant even when subjected to high levels of disturbance. According to Hardesty et al. (1988), the invasion of Croton floribundus in anthropised caatingas is due to the production of seeds in large quantities, which are easily dispersed, both at the moment of fruit dehiscence and afterwards, superseding the mortality rate. In this way, these species can be considered to be one of the main responsible for natural regeneration in the fragment.

Low diversity indices are expected when the sample shows many individuals of the same species and many species with very different numbers of individuals for each (Kent & Coker 1994; Felfili & Rezende 2003).

Table 2 - Floristics of tree species in a caatinga area in the Córrego do Quixinxé/Antônio Conselheiro settlement, municipality of Ocara, Ceará.

FAMILY	SPECIES	NAME POPULAR
BORAGINACEAE	*Auxemma oncocalyx* (Fr.All.) Baill.	Pau-Branco-Black
BURSERACEAE	Commiphora leptophloeos (Mart.)	Imbiratanha
COMBRETACEAE	Combretum leprosum Mart.	Mofumbo
EUFORBIACEAE	Croton floribundus Spreng.	Quince
	Jatropha molíssima (Pohl) Baill.	Jatropha
FABACEAE	Mimosa tenuiflora (Willd.) Poir	Jurema-Preta
	Poincianella bracteosa (Tul.)L.P.Queiroz	Catingueira
	Anadenanthera colubrina var. cebil	

	(Griseb.) Altschul	Angico
	Mimosa caesalpinifolia (Benth.) Ducke	Saviour
	Piptadenia stipulacea (Benth.) Ducke	Jurema-White
OLACACEAE	Ximenia americana L.	Plum
RHAMNACEAE	Ziziphus joazeiro Mart.	Juazeiro

Source: Prepared by the authors

The individuals inventoried corresponded to a total density of 702.0 individuals/ha, a value close to that found in environments that have undergone similar anthropogenic action (PEREIRA et al., 2003; ANDRADE et al., 2005; ARAUJO, 2007), although much lower than most studies in caatinga sensu stricto, which can be explained by the great variability of caatinga physiognomies and/or the differences in anthropogenic pressures exerted. When compared with the results obtained by Costa et al. (2002) in a study carried out in the Seridó region of Rio Grande do Norte, these authors found densities ranging from 3,050 to 4,220 ind. ha-1 and lower densities at 10 other sites (360 to 2,290 ind. ha-1). The high density of the forest fragment studied can be explained as a reflection of the low level of anthropogenic activity over the last 30 years, resulting in greater conservation of the forest remnant. The total basal area was 3.47 m^2/ha, which is considered low, although it is in line with others found in caatinga environments, where values range from 4 to 52 m^2/ha (SAMPAIO, 1996). Araújo et al. (2012) found that the total Basal Area (BA) of the individuals classified in the arboreal shrub layer was 28.77 m^{-2}. ha-1. This result was higher
to the values found by Santana and Souto (2006) (17.50 m-2. ha-1). However, the value was lower than that recorded by Rodal et al. (2008) (34.29 m-2. ha-1) studying a caatinga area.

The species with the highest importance value (VI) was Croton floribundus, which had a density of 453.0 individuals/ha, a frequency of 100% and absolute dominance of 0.92 m^2/ha, which generated an importance value of 112.40, corresponding to 90.85% of the sample (Table 3).

Combretum leprosum was the second species with the highest relative frequency (68%) of representativeness, referring to the percentage of plots in which this species occurs in the area. Other species contributed similarly in terms of their presence in the percentage of plots, such as M. tenuiflora (60%) and P. bracteosa (46%), while the others accounted for less than 22%.

The largest diameter observed was 27.0 cm belonging to an individual of P. bracteosa and the second largest was 23.0 cm corresponding to two individuals of Mimosa tenuiflora and Anadenanthera colubrina. The average diameter was 6.47 cm.

Table 3 - Phytosociological variables analysed for tree species in a caatinga area in the Córrego do Quixinxé/Antônio Conselheiro settlement, municipality of Ocara, Ceará. Where; Nind= number of individuals; DA= absolute density, DR= relative density, DoA= absolute dominance, DoR= relative dominance, FA= absolute frequency, FR= relative frequency, IVI= importance value index and IVC= cover value index.

Species	***DA***	**DR**	**DoA**	**DoR**	**FA**	**FR**	**IVI**	**IVC**
Croton floribundus	453,0	64,53	0,92	26,32	100,0	21,65	112,50	90,85
Mimosa tenuiflora	37,0	5,27	0,81	23,35	60,0	12,99	41,61	28,62
Poincianella bracteosa	28,0	3,99	0,73	20,95	46.0	9,96	34,90	24,94
Combretum leprosum	82,0	11,68	0,17	4,83	68,0	14,72	31,23	16,51
Ximenia americana	26,0	3,70	0,02	0,68	46,0	9,96	14,34	4,38
Auxemma oncocalyx	13,0	1,85	0,16	4,67	22,0	4,76	11,29	6,52
Ziziphus joazeiro	12,0	1,71	0,15	4,37	22,0	4,76	10,84	6,08
Commiphora	10,0	1,4	0,15	4,39	20,0	4,3	10,1	5,8

leptophloeos		2				3	4	1
Anadenanthera colubrine	6,0	0,85	0,21	6,11	12, 0	2,60	9,56	6,96
Mimosa caesalpinifolia	12, 0	1,71	0,09	2,57	22, 0	4,76	9,04	4,28
Piptadenia stipulacea	14, 0	1,99	0,01	0,23	26, 0	5,63	7,85	2,22
Jatropha molíssima	9,0	1,28	0,05	1,53	18, 0	3,90	6,71	2,81

Source: Prepared by the author

The average diameter was considered small and strongly affected by the presence of many shrub species, which normally have small diameter stems. The percentage distribution of the number of individuals per species reflects the low timber volume of this caatinga area in the Córrego do Quixinxé Settlement, even after 5 years of natural regeneration. The sharp rise and fall, respectively, in the number of individuals can be explained by the intense intervention in the natural succession process, more necessarily caused by the clear-cutting carried out in the area in 2009.

In general, it was observed that the majority of individuals in the area were small, i.e. with an average height of 3.20 metres, and that the majority (453 individuals) belonged to the pioneer ecological group (64.53%), which can be inferred that the area is at an early stage of succession, as is the case with Croton floribundus.

It is possible to observe that the area has floristic elements that are distinguished by various factors on different scales, mainly by the effect caused by clear-cutting 5 years ago. Locally, the altitude, depth, chemical composition of the soil, topography and micronutrients have been identified as important in the selection and establishment of species (PAGANO; LEITAO FILHO, 1987; RODRIGUES et al., 1989). Above all, the geographical proximity between the areas and the current states of conservation, attributed to different levels of anthropism, were fundamental to the formation of the groupings.

CHAPTER 5

CONCLUSIONS

In the natural regeneration of the studied area, the Euphorbiaceae and Fabaceae families had the highest number of species.

The species with the highest importance value (VI) was Croton floribundus, which had a density of 453.0 individuals/ha. Combretum leprosum was the second species with a density of 82.0 individuals/ha, followed by M. tenuiflora and P. bracteosa.

The value of the Shannon diversity index (H') was 1.40 and the Pielou Equability index (J') was 0.54.

The high number of individuals with reduced diameter and height points to the existence of a community that is still in the process of recovering its original structure as a result of itinerant farming five years ago;

Croton floribundus was the species that showed the best establishment in the parameters analysed (density, frequency, dominance, importance value and total natural regeneration) and was the only one present in all the plots.

REFERENCES

ALCOFORADO FILHO, F. G.; SAMPAIO, E. V. S. B.; RODAL, M. J. N. Floristics and phytosociology of a remnant deciduous thorny tree vegetation in Caruaru, Pernambuco. Acta Botânica Brasílica, Porto Alegre, v. 17, n. 2, p. 287-303, 2003.

Almeida JR., E. B.; Zickel, C. S.; Pimentel, R. M. M. (2006) Characterisation and biological spectrum of the vegetation of the sandy coast of Rio Grande do Norte. **Rev.Geog. UFPE**. v. 23, n.3, p.45-58.

ALVES, José Jakson Amâncio. Geoecology of the caatinga in the semi-arid Northeast of Brazil. Rio Claro, Climatologia e Estudos da Paisagem,- CLIMEP, v.2, n.1, p. 58-71. 2007.

Araújo, D. S. D. (2000) **Floristic and phytogeographic analysis of the sandbanks of the State of Rio de Janeiro.** Thesis (Doctorate in Ecology). Postgraduate Programme in Ecology, Federal University of Rio de Janeiro.

Araújo, D. S. D.; Henriques, R. P. B. (1984) Floristic analysis of the restingas of the state of Rio de Janeiro. In: Lacerda, L. D.; Araújo, D. S. D.; Cerqueira, R.; Turcq, B. (org.), **Restingas: origem, estrutura e processos.** Niterói - RJ: CEUFF. pp. 159 - 193.

ARAÚJO, E. L,; SILVA, K. A.; FERRAZ. A. M. N.; SAMPAIO, E. V. S. B.; SILVA, S. I. Herbaceous diversity in rocky, flat and riparian microhabitats in a caatinga area, Caruaru, PE, Brazil. Acta Botânica Brasílica. São Paulo, v. 19, n. 2, p. 285-294. 2005.

Calixto Júnior, J. T.; Drumond, M. A.; Alves Júnior, F. T.; Structure and Spatial Distribution of Mimosa tenuiflora (Willd.) Poir. in two **caatinga** fragments in Pernambuco **Revista Caatinga**, vol. 24, núm. 2, abril-junho, 2011, pp. 95-100 Universidade Federal Rural do Semiárido Mossoró, Brazil.

Cantarelli, J. R. R. (2003) **Vascular flora and physiognomic characterisation of a sandbank in the Guadalupe Environmental Protection Area - Pernambuco.** Dissertation (Master's Degree in Botany). Postgraduate Programme in Botany, Federal Rural University of Pernambuco, Recife.

COSTA, T. C. C.; ACCIOLY, L. J. O.; OLIVEIRA, M. A. J.; BURGOS, N.; SILVA, F. H. B. B. Phytomass mapping of the "Seridó caatinga" vegetation by

the plant area and the normalised difference vegetation indices. Scientia Agrícola, v. 59, p. 707-715. 2002.

Felfili, J. M.; Rezende, R. P. (2003) Concepts and methods in phytosociology. **Commun. Técn. Florestais** - UnB, 5(1):1-68.

FERRAZ, E. M. N.; RODAL, M. J. N.; SAMPAIO, E. V. S. B.; PEREIRA, R. C. A. Floristic composition in stretches of caatinga and highland marsh vegetation in the Pajeú Valley region, Pernambuco. Revista Brasileira de Botânica, São Paulo, v. 21, n. 1, p. 7-15, 1998.

FIGUEIRÔA, J.M.; ARAÚJO, E.L.; PAREYN, F.G.C.; CUTLER, D.F.; GASSON, P.; LIMA, K.C.; SANTOS, V.F. Seasonal variations in the survival and biomass production of Caesalpinia pyramidalis Tul. after clear-cutting and implications for the management of the species. **Revista Árvore**, v.32, n.26, p. 1041-1049, 2008.

FREITAS, R.A. C. et al. Floristic and phytosociological study of the shrub extract of two environments in Messias Targino, Rn/Pb border. Revista Verde Mossoró - RN - Brasil, v.2, n.1, p. 135-147 January/July 2007.

FREITAS, R.A.C. et.al. Floristic and phytosociological study of the shrub layer of two environments in Messias Targino, RN/PB border. Green Journal of Agroecology and Sustainable Development. Mossoró, RN v.2, p.135-147, 2007.

Gentry, A. (1988) Changes in plant community Diversity and Floristic composition on environmental and geographical gradients. In: **Annals of the Missouri Botanical Garden**, 35:1 - 34.

HARDESTY, L .H.; BOX, T.W.; MALECHEK, J. C. Season of cutting affects biomass production by coppicing browse species of the Brazilian caatinga. Journal of Range Management, v 41(6): p 477-480.1988.

Kent, M.; Coker, P. (1994) **Vegetation description and analysis - A practical Approach.**

West Sussex: John Wiley & Sons Ltd.

LEITE, V. T. Análise da Estrutura Fitossociologica do Estrato Arbustivo- Arbóreo de duas tipologias de Caatinga ocorrentes no Município de Jucurutu de São João do Cariri-PB. AREIA-PB, Marli 1999, Dissertation (Graduation) UFPB, p. 13-14 and 35.

Lemos, M.C.; Pellens, R. Lemos, L.C. (2001) Profile and floristics of two stretches of coastal forest in the municipality of Maricá - RJ. **Acta Bot. Bras.**15(3):321 - 334.

LIRA, R.B. Floristic Composition and Phytosociological Analysis of the Shrub Component in the Açú National Forest - "FLONA" in the municipality of Assú-RN. 2003. 29p. Monograph (Graduation in Agronomy) - Mossoró School of Agriculture - ESAM, Mossoró.

MIRANDA, I. S.; ACCIOLY, L. J. O.; SILVA, F. H. B. (2000). Vegetation structure of two caatinga areas in the Seridó desertification centre,

Rio Grande do Norte. 264-265p. In: Abstracts of the 51st National Botanical Congress. Botanical Society of Brazil. Brasília-DF.

OLIVEIRA, P. T. B. de.; TROVAO, D. M. de B. M.; CARVALHO, E. C. D. de.; SOUZA, B. C. de.; FERREIRA. L. M. R. Floristics and phytosociology of four vegetation remnants in mountain areas in the Cariri region of Paraíba. Revista Caatinga, Mossoró, v. 22, n. 4, p.169-178, Oct.-Dec. 2009.

REIS, A. M. S. Organisation of the herbaceous layer in a caatinga area of Pernambuco in consecutive years. 2003. 55p. Dissertation (Master's Degree in Plant Botany). Federal Rural University of Pernambuco, Recife, PE. 2003.

RODAL, M.J.N. 1992. **Phytosociology of shrub and tree vegetation in four caatinga**

areas in Pernambuco. Doctoral thesis, State University of Campinas, Campinas.

Sá, C.F.C. (1992) A vegetação da Restinga de Ipitangas, Reserva Ecológica de Jacarepiá, Saquarema (RJ): Fisionomia e lista de Angiospermas. **Arch. J. Bot.** RJ, 21: 87 - 102.

SAMPAIO, E.V.S.B.; ARAÚJO, E.L.; SALCEDO, I.H.; TIESSEN, H. Regeneration of caatinga vegetation after slash and burn, in Serra Talhada, PE. **Pesquisa Agropecuária Brasileira**, v. 33, n.5, p. 621-632, 1998.

Silva, J. G.; Oliveira, A. J. (1989) The restinga vegetation of the municipality of Maricá - RJ. **Acta Bot. Bras.**, 3 (2): 253-272.

TROVAO, D. M. de B. M.; FREIRE, A. M.; MELO, J. I. M. de. Floristics and phytosociology of the woody component of the riparian forest of the Bodocongó stream, semi-arid Paraiba. Revista Caatinga, Mossoró, v. 23, n. 2, p. 78-86, abr.-jun., 2010.

ANNEXES

Photos of the study area

MIX
Papier aus verantwortungsvollen Quellen
Paper from responsible sources
FSC® C105338

Printed by Books on Demand GmbH, Norderstedt / Germany